# Report of a Workshop on Big Data

Committee for Science and Technology Challenges to
U.S. National Security Interests

Division on Engineering and Physical Sciences

# NATIONAL RESEARCH COUNCIL
*OF THE NATIONAL ACADEMIES*

THE NATIONAL ACADEMIES PRESS
Washington, D.C.
www.nap.edu

**THE NATIONAL ACADEMIES PRESS   500 FIFTH STREET, NW   Washington, DC 20001**

NOTICE: The project that is the subject of this report was approved by the Governing Board of the National Research Council, whose members are drawn from the councils of the National Academy of Sciences, the National Academy of Engineering, and the Institute of Medicine. The members of the committee responsible for the report were chosen for their special competences and with regard for appropriate balance.

This study was supported by Contract HHM402-10-D-0036 between the Defense Intelligence Agency and the National Academy of Sciences. Any views or observations expressed in this publication are those of the author(s) and do not necessarily reflect the views of the organizations or agencies that provided support for the project.

International Standard Book Number: 13: 978-0-309-26688-8
International Standard Book Number: 10: 0-309-26688-2

Limited copies of this report are available from the Division on Engineering and Physical Sciences, National Research Council, 500 Fifth Street, NW, Washington, DC 20001; (202) 334-2400.

Additional copies of this report are available from the National Academies Press, 500 Fifth Street, NW, Keck 360, Washington, DC 20001; (800) 624-6242 or (202) 334-3313; http://www.nap.edu.

Copyright 2012 by the National Academy of Sciences. All rights reserved.

Printed in the United States of America

# THE NATIONAL ACADEMIES
*Advisers to the Nation on Science, Engineering, and Medicine*

The **National Academy of Sciences** is a private, nonprofit, self-perpetuating society of distinguished scholars engaged in scientific and engineering research, dedicated to the furtherance of science and technology and to their use for the general welfare. Upon the authority of the charter granted to it by the Congress in 1863, the Academy has a mandate that requires it to advise the federal government on scientific and technical matters. Dr. Ralph J. Cicerone is president of the National Academy of Sciences.

The **National Academy of Engineering** was established in 1964, under the charter of the National Academy of Sciences, as a parallel organization of outstanding engineers. It is autonomous in its administration and in the selection of its members, sharing with the National Academy of Sciences the responsibility for advising the federal government. The National Academy of Engineering also sponsors engineering programs aimed at meeting national needs, encourages education and research, and recognizes the superior achievements of engineers. Dr. Charles M. Vest is president of the National Academy of Engineering.

The **Institute of Medicine** was established in 1970 by the National Academy of Sciences to secure the services of eminent members of appropriate professions in the examination of policy matters pertaining to the health of the public. The Institute acts under the responsibility given to the National Academy of Sciences by its congressional charter to be an adviser to the federal government and, upon its own initiative, to identify issues of medical care, research, and education. Dr. Harvey V. Fineberg is president of the Institute of Medicine.

The **National Research Council** was organized by the National Academy of Sciences in 1916 to associate the broad community of science and technology with the Academy's purposes of furthering knowledge and advising the federal government. Functioning in accordance with general policies determined by the Academy, the Council has become the principal operating agency of both the National Academy of Sciences and the National Academy of Engineering in providing services to the government, the public, and the scientific and engineering communities. The Council is administered jointly by both Academies and the Institute of Medicine. Dr. Ralph J. Cicerone and Dr. Charles M. Vest are chair and vice chair, respectively, of the National Research Council

**www.national-academies.org**

## COMMITTEE FOR SCIENCE AND TECHNOLOGY CHALLENGES TO U.S. NATIONAL SECURITY INTERESTS

J. JEROME HOLTON, Tauri Group, *Chair*
EDWARD M. GREITZER, Massachusetts Institute of Technology, *Vice Chair*
BRIAN BALLARD, APX Labs
KENNETH I. BERNS, University of Florida College of Medicine
ANN N. CAMPBELL, Sandia National Laboratories
DEAN R. COLLINS, Defense Advanced Research Projects Agency (Retired)
SHARON C. GLOTZER, University of Michigan
J.C. HERZ, Batchtags, LLC
KENNETH A. KRESS, KBK Consulting, Inc.
DARRELL LONG, University of California, Santa Cruz
JULIE J.C.H. RYAN, George Washington University
JANET A. THERIANOS, Independent Consultant (USAF, retired)
ELIAS TOWE, Carnegie Mellon University
ALFONSO VELOSA III, Gartner, Inc.
ELI YABLONOVITCH, University of California, Berkeley

*Staff*

TERRY JAGGERS, Lead DEPS Board Director
DANIEL E.J. TALMAGE, JR., Study Director
SARAH CAPOTE, Research Associate
MARGUERITE SCHNEIDER, Administrative Coordinator
DIONNA ALI, Senior Program Assistant
CHRIS JONES, Financial Associate

# Preface

The workshop described in this report is the first in a series of three workshops, held in early 2012 to further the ongoing engagement among the National Research Council's (NRC's) Technology Insight—Gauge, Evaluate, and Review (TIGER) Standing Committee, the scientific and technical intelligence (S&TI) community, and the consumers of S&TI products. A restricted version of this report is available by contacting the Public Affairs Office of the sponsoring agency (Defense Intelligence Agency) directly.

We express our appreciation to the members of the Committee for Science and Technology Challenges to U.S. National Security Interests for their contributions to the development of this report. We are also grateful for the active participation of many members of the technology community in the workshop, as well as to the sponsor for its support. The committee also expresses sincere appreciation for the support and assistance of the NRC staff, including Terry Jaggers, Daniel Talmage, Sarah Capote, Marguerite Schneider, Zeida Patmon, and Dionna Ali.

J. Jerome Holton, *Chair*
Edward Greitzer, *Vice Chair*
Committee for Science and Technology
   Challenges to U.S. National Security Interests

# Acknowledgment of Reviewers

This report has been reviewed in draft form by individuals chosen for their diverse perspectives and technical expertise, in accordance with procedures approved by the National Research Council's (NRC's) Report Review Committee. The purpose of this independent review is to provide candid and critical comments that will assist the institution in making its published report as sound as possible and to ensure that the report meets institutional standards for objectivity, evidence, and responsiveness to the study charge. The review comments and draft manuscript remain confidential to protect the integrity of the deliberative process. We wish to thank the following individuals for their review of this report:

Gilman Louie, Alsop Louie Partners,
David Maddox (NAE), Consultant,
Alton Romig (NAE), Lockheed Martin Aeronautics Company, and
Mikhail Shapiro, University of California, Berkeley.

Although the reviewers listed above have provided many constructive comments and suggestions, they were not asked to endorse the views of individual participants, nor did they see the final draft of the report before its release. The review of this report was overseen by Louis J. Lanzerotti (NAE), New Jersey Institute of Technology. Appointed by the NRC, he was responsible for making certain that an independent examination of this report was carried out in accordance with institutional procedures and that all review comments were carefully considered. Responsibility for the final content of this report rests entirely with the authoring committee and the institution.

# Contents

1 MOTIVATION FOR THE WORKSHOP    1

2 FIRST-DAY PRESENTATIONS    3
   Big Data Analytics, 3
   Discussion of Big Data, 4
      Big Data as Too Much Data, 4
      Big Data as Ubiquitous Sensor Data, 4
      Big Data as Data Fusion Challenges, 5
      Big Data as Too Much of a Good Thing, 5
   Big Data Feeds—1, 5
   Computational Data, 6
   Big Data Feeds—2, 6
   Discussion of Vulnerabilities, 7
   Data Discovery, 7
   Social Networks, 8

3 SECOND-DAY DISCUSSION    9
   Technical, 9
   Temporal, 9
   Personnel, 9
   Blue Process, 10
   Closing Remarks, 10

APPENDIXES
A  Committee Biographies    15
B  Workshop Agenda and Participants    21
C  Speaker Biographies    23

# 1
# Motivation for the Workshop

In 2012, the Defense Intelligence Agency (DIA) approached the National Research Council's TIGER standing committee and asked it to develop a list of workshop topics to explore the impact of emerging science and technology. From the list of topics given to DIA, three were chosen to be developed by the Committee for Science and Technology Challenges to U.S. National Security Interests. The first in a series of three workshops was held on April 23-24, 2012. This report summarizes that first workshop, which explored the phenomenon known as big data.

The objective for the first workshop is given in the statement of task (see Box 1-1), which states, "The workshop will review emerging capabilities in large computational data to include speed, data fusion, use, and commodification of data used in decision making. The workshop will also review the subsequent increase in vulnerabilities over the capabilities gained and the significance to national security."

The committee devised an agenda that helped the committee, sponsors, and workshop attendees probe issues of national security related to so-called big data as well as gain understanding of potential related vulnerabilities. The workshop (see the agenda in Appendix B) was used to gather data that is described in this report, which presents views expressed by individual workshop participants. Although the committee is responsible for the overall quality and accuracy of the report as a record of what transpired at the workshop, the views presented are not necessarily those of all the workshop participants, the committee, or the National Research Council. This workshop report was not intended to provide a comprehensive review of the state of big data.

Chapter 2 of this report summarizes presentations made and discussions held on the first day of the workshop, April 23, 2012. Chapter 3 chronicles the presentations and discussions from the second day of the workshop, April 24, 2012. The three appendixes contain, in order, the biographies of the committee members, the workshop agenda and lists of attendees, and the biographies of the presenters.

> **BOX 1-1**
>
> **Statement of Task**
>
> An ad hoc committee will plan and conduct three workshops on the science and technology (S&T) fields noted below that have potential impact on U.S. national security.
>
> - Big Data—The workshop will review emerging capabilities in large computational data to include speed, data fusion, use, and commodification of data used in decision making. The workshop will also review the subsequent increase in vulnerabilities over the capabilities gained and the significance to national security.
> - Future of Antennas—The workshop will review trends in advanced antenna research and design. The workshop will also review trends in commercial and military use of advanced antennas that enable improved communication, data transfer, soldier health monitoring, and other overt and covert methods of standoff data collection.
> - Future Battlespace Situational Awareness—The workshop will review the technologies that enable battlespace situational awareness 10-20 years into the future for both red and blue forces. The workshop will emphasize the capabilities within air, land, sea, space, and cyberspace.
>
> The committee will design the workshops to address U.S. and foreign research, why S&T applications of technologies in development are important in the context of military capabilities, and what critical scientific breakthroughs are needed to achieve advances in the fields of interest—focusing detailed attention on specific developments in the foregoing fields that might have national security implications for the United States. The workshops will each also consider methodology to track the relevant technology landscape for the future.
>
> Each of the three workshops will feature invited presentations and panelists and include discussions on a selected topic including themes relating to defense warning and surprise. The committee will plan the agenda for the workshops, select and invite speakers and discussants, and moderate the discussions. Each event will result in a workshop summary that will be subject to appropriate institutional review prior to release.

# 2
# First-Day Presentations

## BIG DATA ANALYTICS

*Rod Smith, Emerging Technologies, IBM*

The following is a summary of the presentation made by Rod Smith, who attended the Big Data Workshop via teleconference. According to Smith, IBM recognizes that with the proper tools and understanding of customer needs, there is significant business potential in gleaning information from the burgeoning information loosely termed "big data." Big data grows by the easy sharing of data on the web and with mobile applications, which now is the backbone of social interactions and many business transactions. While the sources of data are growing in both number and magnitude, the cost of storage and processing continues to decline. By utilizing early and direct customer engagement, IBM has developed business insights enabling quick, profitable, and advantageous decisions.

When integrated with data from social media like Twitter (7 terabytes/day) and Facebook (10 terabytes/day), big data acquires a real-time dimension. The value increases even more with the addition of proprietary data. These combined sources of data yield a stronger, discernible signal that illuminates sights and events that might otherwise go unnoticed. One example of the real-time effect of social media is the reporting of the August 2011 earthquake in Mineral, Virginia: Twitter users posted reports in about 40 seconds, whereas the U.S. Geological Survey issued reports, based on seismometer readings, 2 minutes after the same event.

Open source projects like Hadoop[1] and Cassandra[2] are common platforms for big data solutions. The IBM tools for such analyses are refined or modest modifications of rapidly evolving and widely available web technologies so that processing developments using Java, Linux, and XML continue without direct investments from IBM. Using the open source environment is both economical and ensures that efforts continuously remain at the cutting edge of technology.

An example of what IBM can do with these capabilities comes from the support it provided to the Mergers and Acquisition Department of American Express (AMEX), for which IBM produced critical decision information through discovery and analysis of public and private data on intellectual property. The AMEX business question was whether or not the innovation of a specific company was enhanced by a particular business acquisition. The analysis began with a review of the company's patents, ranking them in value in accordance with the number of times they were cited in other patents. This analysis included all U.S. patents (1,400,000) from 2002 to 2009 and another 6,100,000 U.S. and international patents. With approximately one hour of processing time, the analysis showed that one patent had 67 citations, and 24 patents had one citation each. This search triggered an ancillary question, Were these patents involved in litigation?, which resulted in the identification of 3600 cases from the Federal Circuit Court of Appeals (1993-2007).

---

[1]Hadoop is a cross-platform distributed file system that allows massive interaction of computationally independent systems processing enormous amounts of data. It is a product of the Apache Software Foundation.

[2]Cassandra is a product of the Apache Software Foundation, which is dedicated to open source products. Apache Cassandra is a database management system that is designed to handle massive amounts of data distributed across many systems.

IBM believes that such applications are only beginning, demonstrating the start of a possible next wave of business applications. Further, the combination of astute data analytics from IBM and continued contact with customers is key to success. Smith posited that integrating social media analytics is critical to reducing time to value.

## DISCUSSION OF BIG DATA

*Darrell Long and Gilman Louie*

Workshop attendee Gilman Louie and committee member Darrell Long next led a discussion on big data challenges. The challenges of big data are difficult to categorize, primarily because the exact definition of "big data" varies according to the intentions of the speaker. As such, several participants noted that it is important to specify precisely which problem applies in which context and then approach the problem from that definitional space. There was a robust dialog that included four different ways to view the problem: volume[3] of data (too much data), ubiquity of sensor data, data fusion challenges, and too much of a certain type of data.

### Big Data as Too Much Data

Beginning with the problem of big data as simply too much data, or overwhelming amounts of data, many participants felt that the challenge was not new. It was pointed out that too much data had always been a problem, particularly for aggressive collectors of data, as large governments tend to be, and that the result of too much data typically inspires new approaches for handling the increasing amounts of data. Several participants, however, noted two elements of the current era of big data that seemed to be different from previous eras. The first element was the relationship of data to individuals, uniquely and globally; i.e., the big data challenges of this era seem to be mostly about the data associated with individual movements, preferences, sentiments, and thoughts. This situation differs from previous eras in which big data tended to be generated as a result of economic activities, wars, and science. The individualization of big data stems primarily from the social networking phenomenon but is also enabled by the credit and debit card industries and the logistics industry, particularly point-of-sale applications. Other workshop participants noted that the second element was the importance of algorithmic analysis of data; i.e., the use of math, machine learning, and human emotion-behavior analyses (such as sentiment analyses) seems to have made both a quantitative and a qualitative difference in how data is used and interpreted.

### Big Data as Ubiquitous Sensor Data

It was noted by several workshop participants that part of the forcing function of the big data problem is what is called the data ingest challenge: more data is originating from more sensors. These sensors range from social network updates (Facebook posts, tweets, blog posts, etc.) to embedded, distributed utility functions (wifi repeaters, cameras, financial transactions). Some participants suggested that the greater variety of sources of data and data inputs requires different approaches to data integration and analysis, and also contributes to data communication and storage challenges.

---

[3]Some believe that unlike in the "old world" where volume was a problem, in the big data world, volume is a friend: even dirty data can increase the "resolution" of an entity. In the big data world, data is processed differently. Unlike in the old world where data was processed by reducing collections down to semi-finished and finished intelligence (known as the INTs) and then re-integrating it (all-source analysis) to produce knowledge, in the big data world, data is computed all at once and across different data types, to reveal or allow discovery of knowledge and intelligence.

## Big Data as Data Fusion Challenges

The great variety of data types, including audio, video, text, geographic location markers, and photos, were discussed by many workshop participants as having contributed to the growing need for different approaches to data fusion. The point was made that prior to 1999, most of the data available for analysis was structured. Now it is mostly unstructured and varies in terms of format and dimensionality. For example, fusing text to video is challenging, particularly if the video is not annotated in any way that allows the analyst (and the analysis) to know at what point in the video the text is pertinent. It was noted that these types of data fusion issues are of utmost concern to the intelligence community. The fusion problem, of course, is simply the "tip of the iceberg." An issue is how the data can be presented for cognitive review, i.e., how they might be visualized. A workshop participant commented that representational data can contribute to solving this issue, but that may be simply replacing one form of metadata (existing text tags) with a different sort of metadata (representational constructs of the actual data).

## Big Data as Too Much of a Good Thing

There is a point at which conventional approaches to storage (e.g., copies of files) may stop scaling. This was characterized as reflecting the difference between conventional and quantum physics: below the petabyte range, data storage is "Newtonian"; whereas at greater than petabyte sizes, it becomes "Einsteinian." Novel approaches to storage may help, such as the use of mathematical techniques for distributing elements of data sets and then recreating them as needed. Challenges related to representational data versus fully captured data include preprocessing, distribution of processing actions, reduction of communications needs, "data to decision," targeted ads, signatures/signals identification, and "analysis at the edge."

## BIG DATA FEEDS—1

*Eldar Sadikov of Jetlore*

Eldar Sadikov of Jetlore (formerly Qwisper) was asked to present as a representative of the community that exploits large-scale data for social analysis. Jetlore is capable of taking in unstructured data from social networking sites and producing detailed analysis. Sadikov noted that one of the challenges is in natural language processing, particularly in using context to recognize entities and relationships in unstructured texts written in less than grammatically correct language. He discussed how, in contrast to only a few years ago, there is a wealth of data available in addition to the traditional textual content sensor data; these data include geocoordinates, image and video, and data from many other sources that are not evident to the active user. He discussed using this information for non-traditional purposes such as event detection. He referred to the speed of reports generated via Twitter, which are much quicker than those produced through traditional methods. Much of the discussion focused on the profound changes in the amount of data that is available, and the ability to fuse such data to derive information in ways that were not possible in the past. He also discussed where the best work was being done, and he learned that some of the foundational mathematics was done in Russia.

## COMPUTATIONAL DATA

*Chris Gladwin of Cleversafe*

In his presentation on computational data, Chris Gladwin provided an overview on the size and scale of data as technologists move into the next 10 years and the approaches that industry is using to be able to cope with it. According to Gladwin, the world's data requirements are growing exponentially. While much of the data of the past was textual (e.g., HTML 1.0 web pages) or in documents, most of the new data (up to 85 percent of it) is now unstructured data in the form of media (audio, video, images) and other binary large objects (blobs). Complexity arises when these data must be indexed, accessed, and distributed effectively. As an example, Gladwin noted that one of his current customers, Shutterfly, needs to be able to serve 1.4 million photos per hour to its customer base with little to no perceived delay between request and response.

Gladwin noted that the new challenges related to massive amounts of data supporting dynamic requests have given birth over the last decade to a whole new subset of mathematical techniques for storage. Erasure coding for instance, offers a means for breaking data into recoverable chunks and distributing it across more than one storage device, thus hardening the data against loss. Gladwin noted that Cleversafe uses a 20-of-26 strategy in which it needs only 20 slices of data to recover the whole set (the data is stored as 26 slices). He stated, "At a quadrillion bits, you can't assume they are all right." Through error-coding techniques, data can be protected against corruption.

The final portion of the briefing addressed scalability of data storage systems. Gladwin pointed out that Hadoop, for instance, is limited in the number of nodes that are addressable within its name space—limited to a few thousand nodes. Data storage and access needs in the exabyte range would require new computationally efficient ways to build systems capable of supporting these needs. Gladwin noted that his company was able to prove the feasibility of a storage architecture, not yet in use, that could support 10 exabytes of data—the single largest storage architecture in the world. An on-the-spot back-of-the-envelope calculation shows that this highly distributed and decentralized architecture would need roughly 300 million conventional hard drives and consume approximately 2.4 GW just to operate the hard drives (not including the power demands of servers and networking infrastructure for such tasks as thermal management). The discussion then explored the extreme challenge that such a large data storage structure would present to the current ability to generate power at this scale.

## BIG DATA FEEDS—2

*John Marion of Logos Technologies*

John Marion of Logos Technologies described a persistent surveillance system that evolved from a prototype developed at Lawrence Livermore National Laboratory into a series of systems that have been operationally deployed. The fundamental technology is a group of high-resolution image sensors mounted on a gimbaled platform carried aboard aircraft. The challenges presented by such sensors are centered on limited communication bandwidth, and although various communication technologies have been proposed, none has provided the necessary bandwidth. Marion stated that the current strategy is to exchange arrays of magnetic disks physically, but it is understood that techniques need to be developed for processing all the data on the platform in order to reduce the decision latency. By performing computation at the edge, for example, by using models to enable detection of change, Marion said, only small amounts of data need to be transmitted over the limited communications bandwidth that is available. Even though processing at the edge will be necessary, it is not anticipated that raw data will lose its value: some problems can be addressed through processing on the platform, while others will need higher-performance processing and fusion with other data that will be available only on the ground.

## DISCUSSION OF VULNERABILITES

*Al Velosa*

Committee member Al Velosa led a discussion on vulnerabilities. He mentioned three core areas of vulnerability in the big data arena: infrastructure, data and analysis, and tools and technology.

Infrastructure presents opportunity for vulnerabilities. For example, adversaries can and do have the same level of access to equivalent types of infrastructure as the United States does. The infrastructure tends to be built on standardized equipment that is available globally and can be installed by a large number of service providers. For those adversaries who cannot afford the infrastructure, plenty of companies offer this level of infrastructure by means of a "pay as you go" business model. The infrastructure does have the benefit of many data centers with redundant back-ups of the data, but some key facilities can be crippled by disrupting their power supply.

Data and analysis also present a variety of vulnerabilities. Data, and lots of it, is available to opposing forces, often for free. But the proliferation of data, and the speed with which it is used and consumed, sometimes limit how much the United States verify the data. As a result, there is the possibility of false and malicious data being planted in U.S. systems (e.g., false data on stock movements that can drive capital markets or data that can start a panic about a transmittable disease or contamination of food). Data vulnerabilities operate over different time frames. In an attempted manipulation of financial markets, a short-time-frame response might involve an army of bankers immediately figuring out what is happening and then announcing that this is all misinformation. A long-time-frame response might be needed in a scenario where people are faking illness: authorities might need time to discern what is happening and to determine that misinformation has been spread and that there is no cause for alarm. The analysis and communication of the truth associated with these scenarios are also challenging, in that trust becomes a critical issue. Thus data are very susceptible to issues that center on trust.

Tools and technology are widespread and often available on an open source basis. Thus opponents often have access to the same levels of analysis that the United States does. Furthermore, the large number of computer scientists graduating from both U.S. and foreign universities guarantees them a talent base that may develop opportunities and tools that opponents could deny to the U.S.

## DATA DISCOVERY

*Benjamin Reed of Yahoo! Research*

Ben Reed, a research scientist at Yahoo! Research, gave a presentation on data discovery. One of the assumptions on which Yahoo! operates is that everyone has the same kind of infrastructure, as compared to Yahoo!. The secret sauce (what is kept confidential) is the code used to link data pieces. Yahoo! tries to anticipate the information wants of the general online population so that when someone seeks elaboration on a particular piece of news and goes to the Yahoo! website, he or she can easily find those details. For example, Yahoo! kept track of the buzz surrounding the death of Michael Jackson so that people could find out about the details.

Yahoo! also embraced open source implementation (specifically Hadoop). Yahoo! has commoditized hardware, software, and who can use the platforms. Yahoo! data analysis tools are open source (such as Pig[4]) and have contributors from all over the world. Users actually contribute, and do not just use the tools. There is also a Yahoo! Asia office that coordinates these contributions.

---

[4] According to Wikipedia, a "'Pig' is a high-level platform for creating MapReduce programs used with Hadoop. Pig was originally developed at Yahoo! Research around 2006 for researchers to have an ad hoc way of creating and executing map-reduce jobs on very large data sets. In 2007, it was moved into the Apache Software Foundation."

Automated, large-scale data-gathering agents, known as bots (short for software robots), generate a large volume of traffic to Yahoo! and tend to tax Yahoo! with large quantities of queries. Yahoo! deals with bots by giving them a "fake" version of the information they seek. Because attempts to ignore the bot queries, once they are identified as such, simply result in a multiplication of even more bot queries, Yahoo! simply replies with a version of what the bot asked for, minimally satisfying the query, but well enough to pacify the bot and clear bandwidth for other users.

Hardware to process big data is easily accessible, the software is free, and the processing models are accessible, and so big data is no longer a niche market—there is no barrier to the commercial market. During the workshop discussion, a question was asked about whether parallel processing is difficult across multiple nodes with high-performance computing. Yahoo! does do parallel computing, with algorithms designed to solve the big data problem that are often separate and distinct from those ubiquitous to traditional high-performance computing. With big data, a whole lot of information comes in, and not much comes out. In high-performance computing, a little bit of information comes in, but the outputs are tremendous. So, a different type of tool is required for these two data environments.

## SOCIAL NETWORKS

*Paul Twohey of Ness Computing*

Ness Computing (not to be confused with Ness Corporation) is a small start-up headquartered in Los Altos, California. Currently with 15 employees, it embraces data analysis for commercial purposes. The firm's LikeNess search engine, which draws data from various sources of information, such as social networking sites, applies machine learning techniques to tease out patterns that are then used to establish recommendations for users, generating a small profit per transaction. Ness Computing describes what it does in the following terms: "Ness creates products that connect our users with new experiences."

Its flagship product, Ness, is an application that runs on mobile phones to provide users with restaurant recommendations. To seed the analysis, users are asked to input reviews of 10 restaurants. Based on these reviews and the powerful back-end analysis of data from other users and social networks, the Ness "app" provides recommendations on what other restaurants the user can be expected, in all probability, to like.

According to workshop presenter Paul Twohey, his firm's approach to developing products is based on the emerging realities of electronic commerce and free social networking, whereby the user exchanges personal information for services. It requires extensive back-end computing power, which is different from high-performance computing. He stated that the computing approach is that of taking an enormous amount of data, performing complex mathematical analysis (including sentiment analysis), and providing customized output per user. Ness Computing hires only employees with superb math and computer science skills, and a workshop participant voiced that this approach is problematic, given the low availability of individuals with such skills. Several participants at the workshop noted the need for an emphasis by U.S. educational institutions on advanced math skills so that the U.S. workforce can remain competitive in the future.

# 3
# Second-Day Discussion

After the first presentations of the day, workshop participants began several hours of open discourse recapping and further exploring topics raised in the morning and on the previous day. Numerous anecdotes shared demonstrated that "drowning in data" was not a new problem for the intelligence community or for DoD at large, and that the coming paradigm shift arises from the challenge that big data presents more than just a "volume" problem. According to many workshop participants, the big data challenge has at least three aspects—technical, temporal, and personnel—each with very different implications.

## TECHNICAL

The technical aspect of big data encompasses a range of obstacles that hide under the labels of "just hardware" or "just software" or "just human factors." Investment by both the government and the private sector is ongoing in each of these areas, with a growing understanding that the greatest progress lies in attending to all three from a unified perspective rather than treating them as independent investments.

## TEMPORAL

Discussion among the participants revealed two very different time-based challenges, real-time (an increasing rate of data production that accompanied some real-time sensor applications) and retrospective (an increasing amount of data over a larger and larger period of time). Many contributed to the discussion with examples of how the increase in sensor data acquisition rates was making it more and more difficult to transmit in real time to another point for analysis. This reality has prompted a great deal of attention to methods of digesting, parsing, or triaging data, resulting in transmission of only the actionable parts. Alternately, the discussion touched on the exponentially increasing size of historical data sets, which are fueling interest in inference techniques.

## PERSONNEL

Many workshop participants argued that individuals who are trained in and work at the cutting edge of big data are currently in short supply and that the supply is dwindling even as demand continues to grow. Given that a large number of new advanced degrees in this area are awarded to foreign nationals who then return to their countries of origin, some argued that it would seem that efforts to recruit and retain such individuals in positions in the United States should be redoubled.

Some workshop participants said that machines and humans must learn to work together to exploit the burgeoning world of big data. For example, Gary Kasparov's 2005 free-style chess tournament, in which teams could be composed of any combination of humans and computers, was won by two amateur chess players running open source chess engines on simple off-the-shelf laptops—not by grandmasters, prodigies, or chess supercomputers. Big data analysis requires much human interaction and guidance, and the optimal combination is not necessarily the best machines and the brightest humans. It may instead be the right interface between human and machine.

## BLUE PROCESS

### *Asher Sinensky of Palantir Technologies*

Asher Sinensky of Palantir started his talk by using a chess metaphor for human and machine interaction. Humans have an uncanny ability to make decisions and analyze ideas. This ability is unique to humankind. He stated that for the technological enterprise it is important that humans be in the analysis process. Humans are key for conceptualizing new innovations and new ideas after data analysis. See Box 3-1 for Sinensky's full comments.

### *David Thurman of Pacific Northwest National Laboratory*

David Thurman, computing strategy lead at PNNL, asserted that, in the future, computing applications will move toward computer architectures that bring together the different strengths of customized hardware and software capabilities to evaluate different types of distributed data (bringing together many types of computing). A key issue is being able to derive results-generated data from across different agencies.

He said that in 2005 PNNL created a new computer architecture that analyzes data where it resides rather than making copies of the data and transferring it to one central location. This architecture assumes the availability of new highly efficient algorithms tailored to distribute the heterogeneous data sets. Many of these algorithms are derived from the rapidly evolving commercial off-the-shelf (COTS) processing applications of big data.

## CLOSING REMARKS

At the end of the April 2012 workshop, the chair asked committee members and speakers in attendance to make any final comments on what they had heard over the two days. These comments are made as a summary for the workshop:

Ken Kress—"Big data" is more than just a change of scale—it is a more persistent threat than we have previously observed."

Al Velosa—"Progress in the human-machine interface will reduce friction and will allow capability enhancement for the individual, but we will mostly likely experience a fluidity of people more pronounced than we have ever seen."

David Thurman—"I am struck by how different are the threat and impact of big data versus ballistic missiles and other classical threats because of the acceleration of commercially driven offerings, none of which are as controllable as the classical threat domains."

Asher Sinensky—"Now more than at any time in history, we must demand flexibility and adaptability in the tool sets we create for the problem at hand, because those problems are changing faster than ever before, and we don't have time to create a new generation of inflexible tools to counter each new twist."

Mikhail Shapiro—"The highest value should be placed on the human capital, the engineer, and that asset is an asymmetric economic issue."

Brian Ballard—"The big question is how to organize the data and make it accessible to the problem solvers. Cyber is its own category, but big data is a force multiplier of massive scale, with far-reaching implications. Succeeding here will allow us to 'own the net,' delivering advantages that we posit today but even more importantly, advantages of which we are not yet even aware."

## BOX 3-1

## Chess Analogy

*Asher Sinensky*

One of the most important years in the history of big data was 1997, the year that Deep Blue beat Gary Kasparov at chess. At first blush, this might not seem like a big data challenge; chess after all has only 64 spaces, 32 pieces, 6 different types of pieces, and only two players. However, when chess is analyzed more deeply, its true complexity emerges. Claude Elwood Shannon, the so-called father of information theory, showed that the number of legal configurations a chess board could realize is approximately $10^{43}$. This is obviously an enormous number and is sometimes referred to as the Shannon number. A study* by the University of Southern California in early 2011 estimated the world's total digital storage to be on the order of $10^{21}$. In this light, chess is clearly huge when considered against the scale of the digital world. Even beyond that, a Dutch computer scientist, Louis Victor Allis, estimated the game-tree of complexity of chess to be approximately $10^{123}$. That number is roughly 40 orders of magnitude greater than the estimated number of atoms in the entire universe. The act of computationally playing chess is clearly a "big data" problem, and 1997 showed us that computers can do this better than humans can.

The next important year in this story is 2005. In that year, Gary Kasparov decided to host his own chess tournament. In light of Deep Blue, Kasparov become extremely interested in the capabilities of computational systems but also in the ways that computers and humans approach problem solving. Kasparov's 2005 chess tournament was a free-style tournament in which teams could be composed of any combinations of humans and computers available. Grandmasters, prodigies, and chess supercomputers could team up to form super teams. By 2005, it had already been shown that a chess master teamed with a chess supercomputer was far more capable than a supercomputer alone. Computers and humans have different and complementary analytic strengths: computers don't make mistakes, they are highly precise, while humans can use intuition and lateral thinking. These skills can be combined to build truly formidable chess opponents. However, 2005 was different. The winning team, ZackS, performed so well many thought it was actually Kasparov's team. However, the truth was much more intriguing. It turns out that ZackS was actually two amateur chess players running open source chess engines on simple off-the-shelf laptops—no grandmasters, no prodigies, no chess supercomputers.

This was a remarkable outcome that surprised everyone, including Kasparov himself. Kasparov drew the only conclusion he could: "Weak human + machine + better process was superior to a strong computer alone and, more remarkably, superior to a strong human + machine + inferior process." This revelation points to the essential evolution of the conclusion from Deep Blue in 1997—that humans working together with machines can solve big data challenges better than computers alone. Tackling big data means more than just algorithms, high-performance computing, and massive storage—it means leveraging the abilities of the human mind.

---

*See http://www.computerworld.com/s/article/9209158/Scientists_calculate_total_data_stored_to_date_295_exabytes. See also ZackS - http://chessbase.com/newsdetail.asp?newsid=2461; "Friction in Human-Computer Symbiosis: Kasparov on Chess" at http://blog.palantir.com/2010/03/08/friction-in-human-computer-symbiosis-kasparov-on-chess/; and "A Rigorous Friction Model for Human-Computer Symbiosis" at http://blog.palantir.com/2010/06/02/a-rigorous-friction-model-for-human-computer-symbiosis/.

**Appendixes**

# Appendix A
# Committee Biographies

**J. Jerome Holton**, *Chair*, is a senior systems engineer with the Tauri Group, where he supports the BioWatch Systems Program Office within the Office of Health Affairs, Department of Homeland Security (DHS). He provides analysis, advice, and counsel to senior government decision makers on policy, technology, and operations issues related to weapons of mass destruction and their effects on civilian infrastructure, first responders, military forces, and tactical operations. Prior to this, he served in a variety of leadership positions for private-sector companies, spanning the gamut from scientific research start-up to large management consulting firm. Past clients include the Office of the Deputy Assistant to the Secretary of Defense for Counterproliferation and Chemical/Biological Defense, the Chemical Biological Defense Directorate of the Defense Threat Reduction Agency, the Chemical Biological National Security Program of the Department of Energy, and the DHS Science and Technology Directorate. His work extends broadly across the chemical/biological/radiological/nuclear/conventional explosives detection and countermeasures arena. For several years, he focused on the counterproliferation of, counterterrorism/domestic preparedness issues for, and the detection, identification, and decontamination of chemical and biological weapons. Recent accomplishments include fielding information operations tools and enhancing the intelligence, surveillance, and reconnaissance capabilities to detect and defeat improvised explosive devices as well as the development of applique armor solutions to counter explosively formed penetrators. Holton previously served on the NRC's Standing Committee on Defense Intelligence Agency Technology Forecasts and Reviews (TIGER), the Committee for the Symposium on Avoiding Technology Surprise for Tomorrow's Warfighter, and the Committee on Alternative Technologies to Replace Antipersonnel Landmines. He earned his B.S. in physics from Mississippi State University and holds M.S. and Ph.D. degrees in experimental physics from Duke University.

**Edward M. Greitzer (NAE)**, *Vice Chair*, is the H.N. Slater Professor, Department of Aeronautics and Astronautics at Massachusetts Institute of Technology. He received his A.B., S.M., and Ph.D. from Harvard University. Prior to joining MIT in 1977, he was with United Technologies Corporation, and, more recently, he was on leave at United Technologies Research Center as director, Aeromechanical, Chemical, and Fluid Systems. From 1984 to 1996 he was the director of MIT's Gas Turbine Laboratory, and from 1996 to 2002 was associate head, and from 2006 to 2008 deputy head, of the Department of Aeronautics and Astronautics. His research interests have spanned a range of topics in gas turbines, internal flow, turbomachinery, active control of fluid systems, university-industry collaboration, and robust gas turbine engine design; he was the MIT lead for the Cambridge-MIT Institute Silent Aircraft Initiative. He teaches graduate and undergraduate courses in the fields of propulsion, fluid mechanics, thermodynamics, and energy conversion, as well as the department's undergraduate project course. Greitzer is a three-time recipient of the American Society of Mechanical Engineers Gas Turbine Award for outstanding gas turbine paper of the year; in addition, he received the ASME Freeman Scholar Award in Fluids Engineering, the International Gas Turbine Institute Scholar Award, and publication awards from the American Institute of Aeronautics and Astronautics and the Institution of Mechanical Engineers. He has also received the Aircraft Engine Technology Award from the ASME International Gas Turbine Institute, the U.S. Air Force Exceptional Civilian Service Award, and the ASME R. Tom Sawyer Award. He has been a member of the U.S. Air Force Scientific Advisory Board and the NASA Aeronautics Advisory Committee, and he is an Honorary Professor at Beihang University (Beijing). Greitzer has published more than 70 papers and is lead author of the book *Internal Flow: Concepts and Applications*, published by Cambridge University Press. He is a fellow of AIAA and ASME, a member of the National Academy of Engineering, and an International Fellow of the Royal Academy of Engineering.

**Brian Ballard** founded and currently serves as the CEO of APX Labs, a software company focused on leading development into wearable augmented reality products at the nexus of computer vision, user experience, and see-through displays. Previously he served as the director of product development and vice president at Battlefield Telecommunication Systems (BTS), where he led the development of defense-oriented augmented reality and biometric data fusion applications. As part of his portfolio, he was also heavily engaged in developing mobile 3G and 4G networks, devices, and applications for tactical military employments. Prior to joining BTS, Ballard served as the CTO at Mav6, where he was involved in the development of emerging networking and embedded systems technologies for intelligence, surveillance, and reconnaissance (ISR) systems and applications in government and military. He is a highly experienced professional in the field of national intelligence systems and computer engineering. Employed for more than 10 years with the National Security Agency, he has dealt with all forms of data collection, dissemination, processing, and visualization. Ballard holds an M.S. and a B.S. in electrical and computer engineering from Carnegie Mellon University, and a master's of technology management from the University of Maryland. He is currently working on an MBA at the University of Maryland.

**Kenneth I. Berns** (NAS/IOM) is director of the University of Florida Genetics Institute and Distinguished Professor of Molecular Genetics and Microbiology, Medicine. He has served as a member of the Composite Committee of the United States Medical Licensing Examination, chairman of the Association of American Medical Colleges, president of the Association of Medical School Microbiology and Immunology Chairs, president of the American Society for Virology, president of the American Society for Microbiology, and vice-president of the International Union of Microbiological Societies. He is a member of the National Academy of Sciences and the Institute of Medicine. Berns's research examines the molecular basis of replication of the human parvovirus, adeno-associated virus, and the ability of an adeno-associated virus to establish latent infections and be reactivated. His work has helped provide the basis for use of this virus as a vector for gene therapy. Berns's M.D. and his Ph.D. in biochemistry are from the Johns Hopkins University.

**Ann N. Campbell** is director, Information Solutions and Services, at Sandia National Laboratories. Her organization develops and stewards a broad range of software applications and information systems for both internal (enterprise) and external customers to facilitate the delivery of effective national security technologies. At Sandia, she previously served as senior manager and deputy to the chief technology officer for cybersecurity science and technology (S&T). In that role she was responsible for developing and implementing an institutional strategy for cyber S&T. She was recently acting director for Sandia's Cyber Security Strategic Thrust, leading the lab's activities to expand Sandia's cyber workforce and infrastructure, and strategies to provide increased support for Sandia's national security sponsors' cyber missions. Campbell has also served as deputy for technical programs for the Defense Systems and Assessments Strategic Management Unit (DSA SMU). In that role she advised the DSA vice president regarding the

APPENDIX A                                                                                                              17

unit's national security programs, was responsible for strategic planning and the investment strategy for the DSA, and assisted with implementation of the laboratory's cyber strategy. From 2003 to 2007, Campbell led the Assessment Technologies Group in Sandia's Information Systems Analysis Center. She was responsible for development, coordination, and oversight of programs focusing on vulnerability assessments and development of national security solutions in information technologies for multiple government sponsors. From 1999 to 2003 she was manager of the Microsystems Partnerships Department, which assessed and addressed microelectronics vulnerabilities for a variety of government sponsors. In that role Campbell led Sandia's program to support the DoD Anti-Tamper Initiative. She joined the technical staff at Sandia in 1985 and had assignments in the Materials and Process Center and the Microsystems Science, Technology, and Components Center. She conducted research on the microstructure and physical properties of advanced materials, the physics of microelectronics failures, and the development of advanced microelectronics failure analysis techniques. Campbell serves on the National Academies' Standing Committee on Technology Insight–Gauge, Evaluate and Review (TIGER). She is a senior member of IEEE and served as vice president of membership for the IEEE Reliability Society and on the Management Committee and board of directors for the IEEE International Reliability Physics Symposium. She has more than 20 publications and several patents. She holds a B.S. degree in materials engineering from Rensselaer Polytechnic Institute and M.S. and Ph.D. degrees in applied physics (materials science concentration) from Harvard University.

**Dean R. Collins** recently retired as a deputy director of DARPA's Microsystems Technology Office (MTO); as a chief scientist he was responsible for the monitoring, analysis, and evaluation of research projects directed by MTO program managers and also participated in the concept planning for leading MTO into new programs beyond the current state of the art in electronics, photonics, microelectromechanical systems (MEMS), component architectures, and algorithms. He managed the MTO program on integrated circuit cybersecurity. Prior to joining DARPA, Collins was director for the Advanced Research and Development Activity (ARDA) in information technology. ARDA functioned as a joint activity of the intelligence community and the Department of Defense, addressing high-risk/high-payoff information technology problems that had broad impact across both supporting communities. Collins initiated ARDA's key cyber security effort. He was also a member of the intelligence community Advanced Research and Development Committee and managed the ARDA quantum information science effort. Prior to joining ARDA, Collins was with the National Institute of Standards and Technology (NIST), where he was chief of the High Performance Systems and Services Division, the largest division at NIST. This position focused on information technology with a strong commercial bias, and the topics investigated ranged from biometrics to electronic books. Previously, Collins was with Texas Instruments, as director of the System Components Lab, which was responsible for all research on III-V devices, nanoelectronics, photonics, and neural networks. Prior to that, he was director of the Interface Technology Lab, which was responsible for all sensor and display research, including LCDs, DLPs, and CCDs. Collins is a fellow of the IEEE, a member of the American Physical Society, and a registered professional engineer. He has published more than 40 refereed articles and has 10 issued U.S. patents.

**Sharon C. Glotzer** is the Stuart W. Churchill Collegiate Professor of Chemical Engineering and a professor of materials science and engineering at the University of Michigan (UM), Ann Arbor, and is director of research computing in the UM College of Engineering. She also holds faculty appointments in physics, applied physics, and macromolecular science and engineering. She received a B.S. in physics from UCLA and a Ph.D. in physics from Boston University. Prior to joining UM, she worked at the National Institute of Standards and Technology. Her research focuses on computational nanoscience and simulation of soft matter, self-assembly and materials design, and computational science and engineering and is sponsored by the DoD, DoE, NSF, and the J.S. McDonnell Foundation. Glotzer is a fellow of the American Physical Society and of the National Security Science and Engineering Faculty, and she was elected to the American Academy of Arts and Sciences in 2011. She has served on the National Academies' Solid State Sciences Committee; Technology Warning and Surprise study committee; Biomolecular Materials and Processes study committee; Modeling, Simulation, and Games study committee; and Technology Insight–Gauge, Evaluate, and Review (TIGER) Committee. She is involved in roadmapping activities for computational science and engineering, including chairing or co-chairing several workshops, steering committees and pan-agency initiatives, and she serves on the advisory committees for the DOE Office of Advanced Scientific Computing and NSF Directorate for Mathematical and Physical Sciences. Glotzer is also co-founding director of the Virtual School for Computational Science and Engineering under the auspices of the NSF-funded Blue Waters Petascale Computing Project at the National Center for Supercomputing Applications.

**J.C. Herz** is chief executive officer at Batchtags, LLC. She is also a technologist with a background in biological systems and computer game design. Her specialty is massively multiplayer systems that leverage social network effects, whether on the web, mobile devices, or more exotic high-end or grubby low-end hardware. She currently serves as a White House Special Consultant to the Office of the Secretary of Defense (Networks and Information Integration). Defense projects range from aerospace systems to a computer-game-derived interface for next-generation unmanned air systems. Hertz is one of the three co-authors of OSD's Open Technology Development roadmap. She serves on the Federal Advisory Committee for the National Science Foundation's education directorate. In that capacity, she is helping NSF harness emerging technologies to drive U.S. competitiveness in math and science. Hertz was a member of the National Research Council's Committee on IT and Creative Practice and is currently a fellow of Columbia University's American Assembly, where she is on the leadership team of the Assembly's Next Generation Project. In 2002, she was designated a Global Leader for Tomorrow by the World Economic Forum. She is a member of the Global Business Network; a founding member of the IEEE Task Force on Game Technologies; a term member of the Council on Foreign Relations; and a member of the advisory board of Carnegie Mellon's ETC Press. Hertz graduated from Harvard University with a B.A. in biology and environmental studies, magna cum laude. She is the author of two books, *Surfing on the Internet* (Little Brown, 1994), an ethnography of cyberspace before the web, and *Joystick Nation: How Videogames Ate Our Quarters, Won Our Hearts, and Rewired Our Minds* (Little Brown, 1997), a history of videogames which traces the cultural and technological evolution of the first medium that was born digital and how it shaped the minds of a generation weaned on Nintendo. Her books have been translated into seven languages. As a *New York Times* columnist, Hertz published 100 essays on the grammar and syntax of game design between 1998 and 2000. She has also contributed to Esther Dyson's Release 1.0, *Rolling Stone*, *Wired*, *GQ*, and the *Calgary Philatelist*.

**Kenneth A. Kress** is a senior scientist for KBK Consulting, Inc., an affilate of Montana State University's Department of Physics, and a consultant for Booz Allen Hamilton, where he specializes in quantum information science and other technical evaluations and strategic planning for intelligence and defense applications. Some of his past clients include DARPA's Microsytems Technology Office, Noblis, Georgia Tech Research Institute, Mitretek Systems Inc., and Lockheed Martin's Special Programs Division. From 1971 to 1999 he worked in a series of positions at the Central Intelligence Agency's Directorate of Operations, Office of Development and Engineering, and finally, Office of Research and Development (ORD); first as a research and development manager, later as a program manager, and finally as an ORD Office senior scientist responsible for management support, the development of technical and strategic plans, and DOD inter-agency coordination for advanced technology. He is the inventor of the solid-state neutron detector, for which he won an award in 1981. He holds a Ph.D. in physics from Montana State University.

**Darrell D.E. Long** is the Kumar Malavalli Professor of Computer Science at the University of California, Santa Cruz. He holds the Kumar Malavalli Endowed Chair of Storage Systems Research and is director of the Storage Systems Research Center. He received his B.S. in computer science from San Diego State University and his M.S. and Ph.D. from the University of California, San Diego. His dissertation advisor was Jehan-François Pâris. He is a fellow of the Institute of Electrical and Electronics Engineers and of the American Association for the Advancement of Science. He is a member of the IEEE Computer Society, the Association for Computing Machinery, the American Society for Engineering Education, the Usenix Association, Upsilon Pi Epsilon, and Sigma Xi. He has broad research interests in many areas of mathematics and science, and in the area of computer science including data storage systems, operating systems, distributed computing, reliability and fault tolerance, and computer security. His research has been supported by the National Science Foundation; the Department of Energy (Office of Science and National Nuclear Security Administration); Lawrence Livermore, Los Alamos, and Sandia National Laboratories; the Office of Naval Research; and a number of industrial sponsors that include IBM, Microsoft, NetApp, Symantec, LSI Logic, Samsung, Hewlett-Packard, and Data Domain. He served as the vice chair and then chair of the University of California Committee on Research Policy. He has served on the University of California President's Council on the National Laboratories, and on the Science and Technology, National Security, and Intelligence committees. He currently serves on the Science and Technology committee for both Los Alamos and Lawrence Livermore National Laboratories. He previously served on the National Research Council Standing Committee for Technology Insight–Gauge, Evaluate and Review. He continues to serve on numerous committees and advisory panels for various federal government agencies.

**Julie J.C.H. Ryan** is an associate professor and chair of Engineering Management and Systems Engineering at George Washington University. She holds a B.S. degree in humanities from the U.S. Air Force Academy, an M.L.S. in technology from Eastern Michigan University, and a D.Sc. in engineering management from the George Washington University. Ryan began her career as an intelligence officer, serving the U.S. Air Force and the U.S. Defense Intelligence Agency. After leaving government service, she continued to serve U.S. national security interests through positions in industry. Her areas of interest are in information security and information warfare research. She was a member of the National Research Council's Naval Studies Board from 1995 to 1998. She has conducted several research projects and has written several articles and book chapters in her focus area.

**Janet A. Therianos**, a consultant, has 30 years of military experience. She is a U.S. Air Force Academy graduate with an undergraduate degree in aeronautical engineering; an MBA from Harvard Business School; and a masters of arts in air and space power strategy. She was a National Defense fellow and has executive education from Harvard's Kennedy School of government, the Center for Creative Leadership, and the Intelligence Community Senior Leader Program. Therianos has flown several military aircraft and has served as a command pilot, flight examiner, flight instructor, and functional check pilot. She also holds an FAA Airline Transport Pilot rating. Her military career was grounded in operations, but she also had extensive higher-headquarters staff duties, including serving as senior military assistant to the Secretary of the Air Force. Her leadership experiences were threaded throughout her career, including several Commands. Her final military assignment was leading the Air Mobility Command's Directorate of Intelligence, where she was responsible for organizing, training, and equipping the Air Force's global mobility intelligence units. Operationally she led the Command's daily Threat Working Group, which assessed threat levels for all global mobility flight operations.

**Elias Towe** is currently a professor of electrical and computer engineering and the Albert and Ethel Grobstein Professor of Materials Science and Engineering at Carnegie Mellon University. He was educated at the Massachusetts Institute of Technology (MIT), where he received B.S, M.S., and Ph.D. degrees from the Department of Electrical Engineering and Computer Science. Towe was a Vinton Hayes Fellow at MIT. After leaving MIT he became a professor of electrical and computer engineering, and engineering physics at the University of Virginia. He also served as a program manager in the Microsystems Technology Office at the Defense Advanced Research Projects Agency (DARPA) while he was a professor at the University of Virginia. In 2001, he joined the faculty at Carnegie Mellon University. Towe is a recipient of several awards and honors that include the National Science Foundation Young Investigator Award, the Young Faculty Teaching Award, and an Outstanding Achievement Award from the Office of the Secretary of Defense. He is a fellow of the Institute of Electrical and Electronics Engineers (IEEE), the Optical Society of America (OSA), the American Physical Society (APS), and the American Association for the Advancement of Science (AAAS).

**Alfonso Velosa III** is research director for Gartner with a focus on sustainability, business ecosystems, and smart cities. He is also agenda manager for electronic equipment research at Gartner, concentrating on electronics and semiconductor supply chain research, with a particular focus on global trends for manufacturing, consumption, financing, and the key vendors in the market. Velosa has also written extensively about electronics, outsourcing of electronics manufacturing, electronic manufacturing services (EMS), original design manufacturing (ODM), and semiconductor consumption. He previously worked at or consulted for Intel Corporation, NASA Lewis Research Center and NASA Headquarters, Mars & Co., and IBM Research. Velosa graduated from Columbia University with a B.S. in materials science engineering; from Rensselaer Polytechnic Institute with an M.S. in materials science engineering; and from Thunderbird, the Garvin School of International Management, with an M.I.M. in international management.

**Eli Yablonovitch** (NAS/NAE) is an adjunct professor of electrical engineering at UCLA after having served as a full faculty member until 2007. He is currently a professor of electrical and computer engineering at University of California, Berkeley. He graduated with a Ph.D. in applied physics from Harvard University, worked for 2 years at Bell Telephone Laboratories, and then became a professor of applied physics at Harvard. In 1979 he joined Exxon to do research on photovoltaic solar energy; in 1984, joined Bell Communications Research, where he was a Distinguished Member of Staff and also director of Solid-State Physics Research; and in 1992, joined the University of California, Los Angeles, where he became the Northrop Grumman Opto-Electronics Chair and a professor of electrical engineering. Yablonovitch's work has covered a broad variety of topics: nonlinear optics, laser-plasma interaction, infrared laser chemistry, photovoltaic energy conversion, strained-quantum-well lasers, and chemical modification of semiconductor surfaces. Yablonovitch's research focuses on optoelectronics, high-speed optical communications, high-efficiency light-emitting diodes and nanocavity lasers, photonic crystals at optical and microwave frequencies, and quantum computing and communication.

# Appendix B
# Workshop Agenda and Participants

## AGENDA

**April 23-24, 2012**
**The Beckman Center of the National Academies**
**Irvine, California**

**Data as a Commodity**
Rod Smith, Vice President, Emerging Internet Technologies
IBM

**Big Data (Discussion)**
Darrell Long, Kumar Malavalli Professor of Computer Science, and Kumar Malavalli Endowed
Chair of Storage Systems Research Center
University of California, Santa Cruz

**Computational Data**
Chris Gladwin, CEO
Cleversafe

**Big Data Feeds**
Eldar Sadikov, CEO/Founder
Jetlore

John Marion, Director, Persistent Surveillance
Logos Technologies

**Data Discovery**
Benjamin Reed, Research Scientist
Yahoo! Research

**Social Networks**
Paul Twohey, Co-Founder and Vice President
Ness Computing

**Blue Process**
Asher Sinensky, Lead Engineer
Palantir Technologies

David Thurman, Computing Strategy Lead
Pacific Northwest National Laboratory

## PARTICIPANTS

### Committee

Brian Ballard, APX Labs
Kenneth Kress, KBK Consulting, Inc.
Darrell Long, University of California, Santa Cruz
Julie Ryan, George Washington University
Alfonso Velosa, Gartner, Inc.

### Speakers

Chris Gladwin, Cleversafe
John Marion, Logos Technologies
Benjamin Reed, Yahoo! Research
Eldar Sadikov, Jetlore
Asher Sinensky, Palantir Technologies
Rod Smith, IBM
David Thurman, Pacific Northwest National Labor
Paul Twohey, Ness Computing

### Staff

Terry Jaggers, Lead Board Director
Daniel Talmage, Study Director
Sarah Capote, Research Associate
Dionna Ali, Senior Program Assistant

### Guest or Agency Represented

Gilman Louie, Alsop-Louie Partners
Mikhail Shapiro, University of California, Berkeley
Defense Intelligence Agency
Department of Defense
National Geospatial Intelligence Agency
United States Air Force

# Appendix C
# Speaker Biographies

**Chris Gladwin** founded Cleversafe in 2004. Previously he was the creator of the first workgroup storage server at Zenith Data Systems and was a manager of corporate storage standards at Lockheed Martin. Gladwin also created and managed a number of successful new technology start-ups, including MusicNow, which was acquired by Circuit City. He has been the creative force behind the development of the first dispersed storage system to solve the growing global problem of big data storage. Gladwin understood the growing issues surrounding unstructured data and the inability of traditional technology solutions to accommodate the explosive growth of digital assets such as audio, video, and imaging. He applied advances in dispersed information technology to storage to create a reliable, cost-effective, secure solution with a limitless ability to scale. Gladwin holds a degree in engineering from the Massachusetts Institute of Technology.

**John Marion** has a broad background in lasers and optical materials, strategic defense, and persistent surveillance. At Lawrence Livermore National Laboratory he led the team that pioneered the field of persistent surveillance, developing hardware and image-processing capabilities and field testing of these new systems. At Logos he led the technical effort to develop a deployable system, resulting in Constant Hawk, the first persistent surveillance system in-theater, as well as championing the use of this new intelligence collection and exploitation paradigm in the DoD and the intelligence community. Currently he leads the Logos team developing the Kestrel persistent surveillance systems for aerostat deployment and is developing future persistent surveillance systems, including the analysis and visualization tool development for exploitation of the imagery. His group also develops novel systems for the intelligence community and is leading a cyber defense technology development effort sponsored by DARPA. He provides technical analysis and support to NAVAIR, ARL, NVESD, DARPA and the CIA.

**Benjamin Reed** is a research scientist at Yahoo! Research. He has worked for almost 2 decades in industry, in positions ranging from work as an intern on cad/cam systems, to shipping and receiving applications in OS/2, AIX, and CICS, to operations, to system admininistration research and Java frameworks at IBM Almaden Research (11 years). He arrived at Yahoo! Research 3 years ago to work on the largest distributed-computing problems. His main interests now are large-scale processing environments and highly available and scalable systems. He has worked largely in open source, including writing and maintaining the Linux Aironet wireless driver. His research project at IBM grew into OSGI which is now in application servers, cars, and mobile phones. Two projects for which he has led research are Pig and ZooKeeper, which are Apache Software Foundation projects.

**Eldar Sadikov** is on leave from the Ph.D. program in computer science at Stanford University. While at Stanford, Eldar conducted research in web search and social network mining. He also worked at Google and at Microsoft Research in web search. He founded Qwhisper, whose goal is to make exponentially growing social content more useful by automatically inferring its meaning and giving it structure. Qwhisper is a highly stimulating intellectual environment in which robust distributed systems are built that handle web-scale data and design algorithms that challenge the published state of the art.

**Asher Sinensky** holds a Ph.D. from MIT in materials science and engineering and oversees product development at Palantir Technologies, where he is directly responsible for plotting the development roadmap for all features and functions. He has been involved in national security for a decade including having worked at Sandia and Lawrence Livermore National Laboratories on projects related to bio security and detection of chemical and biological pathogens. He has received several security-related awards, including the Sandia National Security Fellowship and a Department of Homeland Security Fellowship. At MIT he explored techniques in nanoscale detection of organic molecules such as anthrax DNA. Sinensky is involved in numerous Palantir Technologes deployments across the defense, intelligence community, and law enforcement spaces.

**Rod Smith** is an IBM fellow and vice president of the IBM Emerging Internet Technologies organization, where he leads a group of highly technical innovators who are developing solutions to help organizations realize the value of big data. His early advocacy in the industry has played an important role in the adoption of technologies such as J2EE, Linux, Web services, XML, rich Internet applications, and various wireless standards. As an IBM fellow, Smith is helping lead IBM's efforts around big data analytics and the application of IBM Watson-like technologies to business solutions, helping companies make better decisions more quickly for improved business outcomes. His early identification of emerging technologies has led to a sustained record of achievement in the global software community. Smith has authored numerous invention patents and disclosures, and he is the recipient of several prestigious awards, including the TJ Watson Design Excellence Award. Smith is a computer science graduate of Western Michigan University, and holds an M.A. and a B.A. in economics with a concentration in math from Western Michigan University.

**David Thurman** currently leads computing strategy development at Pacific Northwest National Laboratory (PNNL) as well as providing oversight for activities at the Seattle Research Center in nonproliferation policy analysis, systems engineering, and human-centered analytics. He was previously responsible for program management for PNNL's information analysis portfolio in the national security domain, coordinating a range of research and development projects focused on delivering new analytic capability to a range of government users. With more than 25 years of professional experience in research and university settings, Thurman has managed a variety of information analysis projects that developed new analytic methods and capabilities for a range of client organizations. He previously conducted research on advanced knowledge representation techniques to support intelligence analysis, led efforts to define information integration architectures for the U.S. Department of Homeland Security, studied information analysis methods at the International Atomic Energy Agency, and developed integrated analysis systems for a variety of government clients. Internally at PNNL, he has served in leadership roles for research initiatives on data-intensive computing, threat anticipation, and signature discovery. He is currently leading the definition of a new research initiative in distributed analytics for multi-source data. Thurman was previously a research engineer at Georgia Institute of Technology's Center for Human-Machine Systems Research, developing human interfaces, training systems, and automation in the domains of satellite ground control and commercial aviation. Prior to that, he worked as a software developer in PNNL's Computational Sciences Department, developing advanced data analysis and visualization applications. He has more than 40 peer-reviewed publications and technical reports on a range of information processing and analysis topics. Thurman was a Presidential Fellow at Georgia Institute of Technology, where he received an M.S. in human-machine systems engineering. He also holds B.S. degrees in mathematics and computer science from the University of Oregon. He is a member of IEEE, and ACM and was previously a fellow of the World Affairs Council in Seattle.

**Paul Twohey** is the vice president and a co-founder of Ness Computing. Previously, he was the co-founder of Good Ga.me and also worked as a software engineer for Palantir Technologies. Ness, whose mission is to make search personal, is sometimes referred to as the "Palantir for fun." Twohey gained his M.S. in computer science from Stanford University and his B.S. in electrical engineering and computer sciences from the University of California, Berkeley. He is also the recipient of the 2002 William Everitt Award for Excellence.